From DNA to

Introduction

Maria Rossi

Author: Maria Rossi

Title: From DNA to Diversity: A Student's Introduction to Human Genetics

This book is a self-published work by the author Maria Rossi

ISBN:

TABLE OF CONTENTS

Chapter 1: Introduction to Human Genetics

The Importance of Genetics

Understanding genetics is crucial in unraveling the mysteries of life and our own existence. Genetics is the branch of biology that explores how traits are passed from one generation to another. It is the study of genes, heredity, and the variation of organisms. In the book "From DNA to Diversity: A Student's Introduction to Human Genetics," we aim to provide students with a comprehensive understanding of genetics and its significance.

Genetics holds the key to understanding why we look the way we do, why we may be prone to certain diseases, and how our bodies function. By studying genetics, students can gain insights into their own personal health and make informed choices about their lifestyles. For instance, if they discover that they have a genetic predisposition to certain conditions, they can take preventive measures to minimize the risk.

Moreover, genetics plays a crucial role in the development of new medical treatments and therapies. Scientists are constantly exploring the genetic makeup of diseases to identify potential targets for drugs and therapies. By understanding the genetic basis of diseases, researchers can develop personalized treatments that are tailored to an individual's unique genetic profile. This approach, known as precision medicine, has revolutionized the field of healthcare and has the potential to enhance patient outcomes.

Genetics also contributes significantly to various fields, such as agriculture and conservation. In agriculture, genetic engineering has allowed scientists to develop crops that are resistant to pests and diseases, thus increasing yields and ensuring food security. In conservation, genetics is used to study endangered species and develop strategies for their protection and preservation.

In addition to practical applications, genetics also fuels our curiosity about the natural world. It helps us understand the intricate web of life and the interconnectedness of all living organisms. By studying genetics, students can gain a deeper appreciation for the diversity and complexity of life on Earth.

In "From DNA to Diversity: A Student's Introduction to Human Genetics," we present genetics in an engaging and accessible manner. We delve into the history of genetics, explore the fundamental concepts, and showcase real-world examples. By the end of this chapter, students will have a solid foundation in genetics and will appreciate its importance in various aspects of their lives.

Overall, genetics is a fascinating field that holds immense importance for students. Whether they aspire to become doctors, researchers, or simply want to understand themselves better, studying genetics is a crucial step on their journey to unravel the mysteries of life.

Definition of Human Genetics

Human genetics is a fascinating field of study that explores the genetic makeup and inheritance patterns of human beings. It delves into the complexities of our DNA, genes, and chromosomes, and how they contribute to the diversity we observe in the human population. This subchapter aims to provide students with a comprehensive understanding of the definition of human genetics, its significance, and its role in shaping who we are as individuals and as a species.

At its core, human genetics involves the study of the hereditary information contained within our cells. The genetic material, known as deoxyribonucleic acid (DNA), is composed of nucleotides that form a unique sequence, encoding the instructions for building and maintaining the human body. These instructions are organized into genes, which are functional units of DNA responsible for producing proteins that carry out specific functions in our bodies.

Human genetics also explores the patterns of inheritance and how traits are passed from one generation to the next. This includes understanding the role of genes and the interplay between nature and nurture in shaping our physical characteristics, behavior, and susceptibility to diseases. By studying human genetics, we can unravel the genetic basis of various traits and disorders, such as eye color, height, intelligence, and genetic diseases like cystic fibrosis and sickle cell anemia.

Furthermore, human genetics investigates the diversity within the human population. It examines how genetic variations arise and spread through populations over time, leading to the different races, ethnicities, and individual differences we see today. Understanding this genetic diversity is crucial for addressing questions related to human evolution, migration, and the impact of genetic factors on health and disease.

In conclusion, human genetics is the study of the genetic material and inheritance patterns that shape the human species. It encompasses the exploration of DNA, genes, and chromosomes, as well as the understanding of inheritance, genetic diversity, and the impact of genetic factors on human health. By unraveling the complexities of human genetics, students can gain valuable insights into the fundamental building blocks of life and the remarkable diversity that exists within humanity.

Historical Background

In the captivating world of genetics, understanding the historical background is essential to comprehend the incredible advancements made in this field. This subchapter delves into the rich history of human genetics, providing students with a solid foundation to build upon.

Human genetics as a scientific discipline traces its roots back to the mid-19th century, with the groundbreaking work of Gregor Mendel. Mendel, an Austrian monk, conducted extensive experiments on pea plants, discovering the fundamental principles of inheritance. His findings laid the groundwork for our understanding of genetic inheritance, leading to the birth of the field of genetics.

The early 20th century witnessed remarkable breakthroughs in our understanding of genetics. One of the key milestones was the identification of DNA as the carrier of genetic information by Friedrich Miescher in 1869. This discovery paved the way for further investigations into the structure and function of DNA.

In the 1950s, James Watson and Francis Crick unveiled the remarkable double-helix structure of DNA, a discovery that revolutionized the field of genetics. This breakthrough laid the foundation for the understanding of how genetic information is stored and transmitted from one generation to another.

Another significant milestone was the completion of the Human Genome Project in 2003. This international collaborative effort aimed to decipher the entire human genetic code, unraveling the sequence of genes that make up our DNA. The project provided an invaluable resource for researchers, enabling them to explore the genetic basis of various diseases and traits.

Today, the field of genetics has expanded exponentially, with advancements in technology and research methodologies. Geneticists now have the tools to study the intricacies of our genetic makeup, including the identification of specific genes associated with diseases, the exploration of genetic variation among populations, and the understanding of how genes interact with the environment.

Understanding the historical background of human genetics is crucial for students embarking on a journey into this fascinating field. It provides a context for the incredible discoveries and advancements made over the years, inspiring students to explore the frontiers of genetic research and its potential applications in medicine, agriculture, and beyond.

Key Scientists in Human Genetics

In the vast field of human genetics, several groundbreaking scientists have made significant contributions, paving the way for our current understanding of the complex genetic makeup that shapes human life. These individuals have tirelessly pursued scientific discovery, unraveling the mysteries of our DNA, and revolutionizing the field of genetics. In this subchapter, we will explore the lives and achievements of some of the key scientists in human genetics.

Gregor Mendel, an Augustinian monk, is often referred to as the father of modern genetics. In the mid-19th century, through his meticulous experiments with pea plants, Mendel discovered the fundamental principles of inheritance, introducing the concept of dominant and recessive genes. His work laid the foundation for the study of genetics and inheritance patterns, revolutionizing our understanding of how traits are passed down from one generation to the next.

Building upon Mendel's work, Thomas Hunt Morgan conducted groundbreaking research on the inheritance of traits in fruit flies during the early 20th century. Morgan's experiments led to the discovery of sex-linked traits and the existence of chromosomes, providing evidence that genes are located on these structures. His work on fruit flies allowed scientists to study genetics with greater precision, opening up new avenues for research.

Another pivotal figure in the field of human genetics is James Watson, along with Francis Crick, who unraveled the structure of DNA in 1953. Their discovery of the double helix structure of DNA revolutionized our understanding of how genetic information is stored and transmitted. Watson and Crick's groundbreaking work earned them the Nobel Prize in Physiology or Medicine in 1962.

Moving into the modern era of genetics, we have Jennifer Doudna and Emmanuelle Charpentier, who pioneered the revolutionary gene-editing technique known as CRISPR-Cas9. This technology allows scientists to precisely edit genes, opening up a world of possibilities for treating genetic diseases and improving crop yields. Their groundbreaking work has earned them numerous accolades and has the potential to transform the field of genetics.

These are just a few of the key scientists who have shaped the field of human genetics. Their tireless dedication and groundbreaking discoveries have allowed us to unravel the intricate complexities of our genetic makeup, leading to advancements in medicine, agriculture, and our overall understanding of life itself. As students of genetics, it is essential to recognize and appreciate the contributions of these remarkable individuals who have paved the way for our current understanding of the genetic blueprint that defines us all.

Chapter 2: Basic Concepts in Genetics

Cells and DNA

In the vast world of genetics, the foundation of life lies within cells and DNA. As students delving into the fascinating realm of human genetics, it is crucial to understand the fundamental concepts of cells and DNA, as they form the building blocks of our genetic makeup.

Cells, the basic units of life, are the microscopic powerhouses that make up every living organism. They come in various shapes and sizes, each with specific functions and roles within our bodies. From the intricate neurons in our brain to the mighty muscle cells in our hearts, cells work harmoniously to ensure our bodies function optimally.

At the core of every cell lies the DNA, or deoxyribonucleic acid. DNA can be thought of as the instruction manual for life, containing all the information needed to develop and maintain an organism. It is a complex molecule made up of long chains of nucleotides, which are composed of four bases: adenine (A), thymine (T), cytosine (C), and guanine (G). These bases pair up in a specific manner – A with T, and C with G – forming the famous double helix structure.

DNA carries the genetic code that determines our unique traits, such as eye color, hair texture, and predisposition to certain diseases. It achieves this by encoding genes, which are specific segments of DNA that contain instructions for producing proteins. Proteins, in turn, carry out essential functions in our bodies, such as catalyzing chemical reactions, transporting molecules, and providing structural support.

Understanding the structure and function of DNA has revolutionized the field of genetics. Scientists can now study and manipulate DNA to gain insights into the causes of genetic disorders, develop new treatments, and even explore the mysteries of our evolutionary history.

As students of genetics, it is crucial to grasp the significance of cells and DNA in our everyday lives. By unraveling the mysteries hidden within these microscopic entities, we gain a deeper understanding of ourselves and the incredible diversity that exists within the human population.

So, let us embark on this exciting journey of discovery, as we delve into the intricate world of cells and DNA. Through this knowledge, we can unlock the secrets of life, uncover the origins of our traits, and pave the way for a future where genetic information holds the key to personalized medicine and a deeper understanding of our place in the biological tapestry.

Genes and Chromosomes

In the realm of genetics, the study of genes and chromosomes is crucial to understanding the fundamental building blocks of life. Genes, the hereditary units that determine traits and characteristics, are contained within structures called chromosomes. This subchapter will delve into the intricate world of genes and chromosomes, providing students with a comprehensive introduction to this fascinating field of study.

To begin, let's explore the structure of chromosomes. Chromosomes are thread-like structures found within the nucleus of every cell in our bodies. They are composed of DNA, or deoxyribonucleic acid, which carries the genetic information necessary for the development and functioning of all living organisms. Humans typically have 23 pairs of chromosomes, resulting in a total of 46 chromosomes in each cell.

Each chromosome contains numerous genes, which are specific segments of DNA that provide instructions for the production of proteins. Proteins are essential for the proper functioning of cells and play a crucial role in determining an individual's traits, such as eye color, height, and susceptibility to certain diseases.

The arrangement and organization of genes on chromosomes are intricately structured. Genes are lined up in a linear fashion, with each gene occupying a specific location, or locus, on a chromosome. The different versions of a gene that can occupy a particular locus are known as alleles. These alleles contribute to the genetic variation observed among individuals.

Understanding the relationship between genes and chromosomes is vital in comprehending how genetic traits are inherited. During reproduction, each parent contributes one set of chromosomes to their offspring. This process ensures that offspring inherit a combination of genes from both parents, leading to a unique blend of genetic traits.

Moreover, the study of genes and chromosomes has far-reaching implications. It allows us to understand the underlying causes of genetic disorders, such as Down syndrome, cystic fibrosis, and sickle cell anemia. By unraveling the intricate interactions between genes and chromosomes, scientists can also explore the genetic basis of complex diseases like cancer and develop targeted therapies.

In conclusion, genes and chromosomes are the building blocks of life, providing the instructions necessary for the development and functioning of all living organisms. A thorough understanding of their structure and function is essential for comprehending the inheritance of traits and the underlying causes of genetic disorders. By delving into the world of genes and chromosomes, students can gain a profound appreciation for the intricate workings of genetics and its profound impact on human diversity and health.

The Central Dogma of Genetics

In the fascinating world of genetics, there is a fundamental concept that governs the way genetic information flows and is translated into traits and characteristics. This concept is known as the Central Dogma of Genetics. Understanding this principle is crucial for students delving into the field of genetics, as it forms the foundation for comprehending the complex mechanisms that drive inheritance and evolution.

The Central Dogma can be summarized as the process through which genetic information is transferred from DNA to RNA and then translated into proteins. It outlines the flow of genetic information within living organisms, explaining how the instructions encoded in DNA are ultimately expressed as the physical and biochemical traits we observe.

At the core of the Central Dogma is DNA, the molecule responsible for carrying the genetic information in all living organisms. DNA is made up of a sequence of nucleotides, which act as the building blocks of genetic code. Through a process called transcription, DNA is first copied into a molecule known as messenger RNA (mRNA). This mRNA then carries the genetic information from the nucleus, where DNA resides, to the cytoplasm, where protein synthesis occurs.

Once in the cytoplasm, the mRNA is utilized as a template for translation, the process by which proteins are synthesized. Transfer RNA (tRNA) molecules recognize specific sequences on the mRNA and bring the corresponding amino acids, which are the building blocks of proteins. As the tRNA molecules line up along the mRNA, the amino acids are linked together to form a protein chain.

The Central Dogma provides a framework for understanding how genetic information is transmitted from one generation to the next. When cells divide, DNA is replicated, ensuring that each new cell receives a complete set of genetic instructions. During sexual reproduction, the fusion of sperm and egg cells combines genetic material from two individuals, leading to the generation of offspring with unique genetic characteristics.

By comprehending the Central Dogma, students gain insight into the fundamental processes that underlie genetic inheritance and variation. It provides a roadmap for exploring the intricate mechanisms that drive the evolution of species and the development of human traits. Understanding this concept lays the groundwork for further exploration into the fascinating world of genetics, where the secrets of life's diversity are waiting to be unraveled.

Genetic Variation and Inheritance

Introduction:

Welcome to the subchapter on genetic variation and inheritance! In this section, we will explore the fascinating world of genetics and how it shapes our diversity as humans. Understanding genetic variation and inheritance is crucial in comprehending the complex mechanisms that contribute to the traits and characteristics we possess.

Exploring Genetic Variation:

Genetic variation refers to the diversity in genetic material found within a species. It is the driving force behind the uniqueness of individuals and plays a vital role in evolution. Genetic variation arises from various factors, including mutations, recombination, and gene flow between populations. These processes introduce new genetic material and allow for the accumulation of differences over generations.

The Role of DNA:

DNA, the famous double helix molecule, is the carrier of genetic information. It contains genes, which are segments of DNA that code for specific traits. Genes come in pairs, with one inherited from each parent. The combination of genes inherited from our parents contributes to our individual genetic makeup, known as our genotype.

Inheritance Patterns:

Inheritance patterns determine how traits are passed from one generation to the next. The three main inheritance patterns are dominant, recessive, and co-dominant. Dominant traits are expressed when only one copy of the gene is present, while recessive traits require two copies. Co-dominant traits result in a combination of both gene copies being expressed.

Mendelian Genetics:

Gregor Mendel, an Austrian monk, laid the foundation for the study of genetics through his experiments with pea plants. His work established the principles of inheritance, such as the law of segregation and the law of independent assortment. These laws explain how genes are passed from parents to offspring and how different traits can be inherited independently of each other.

Genetic Disorders:

Genetic variation can also lead to the development of genetic disorders. These disorders are caused by mutations in specific genes that disrupt normal biological processes. Some genetic disorders are inherited in a recessive or dominant manner, while others are caused by spontaneous mutations.

Conclusion:

Genetic variation and inheritance are fundamental concepts in human genetics. By studying these topics, we gain insights into the mechanisms that shape our individuality and the diversity of the human population. Understanding how genes are inherited and the role they play in the development of genetic disorders is crucial in the field of genetics. As students, delving into these concepts will provide a solid foundation for further exploration in the exciting world of genetics.

Chapter 3: Mendelian Genetics

Gregor Mendel and His Experiments

Gregor Mendel, an Augustinian monk, is often referred to as the "Father of Genetics" for his groundbreaking work on pea plants in the mid-19th century. His experiments laid the foundation for our understanding of heredity and paved the way for future advances in the field of genetics.

Mendel's curiosity about how traits are passed from one generation to another led him to conduct a series of experiments with pea plants. He carefully selected different varieties of peas, each with distinct traits such as flower color, seed shape, and plant height. By cross-breeding these plants, Mendel was able to observe how traits were inherited and determine patterns of inheritance.

One of Mendel's most significant discoveries was the concept of dominant and recessive traits. He found that some traits, such as tall plant height, were dominant and always expressed in the offspring when present. On the other hand, traits like short plant height were recessive and only expressed when both parents carried the recessive allele.

Mendel also discovered the principle of segregation, which states that each parent contributes one copy of each gene to their offspring. This principle explained why some traits seemed to disappear in one generation but reappeared in later generations.

Furthermore, Mendel's experiments supported the idea of independent assortment. He observed that traits were inherited independently of each other, meaning that the inheritance of one trait did not influence the inheritance of another. This concept revolutionized our understanding of inheritance and formed the basis for the development of Punnett squares, which are still used today to predict the probability of inheriting certain traits.

Mendel's work went largely unnoticed during his lifetime, but his groundbreaking experiments were rediscovered and recognized by the scientific community several decades later. His discoveries laid the groundwork for modern genetics and greatly influenced the field of human genetics.

Understanding Mendel's experiments and the principles he discovered is crucial for students studying genetics. His work not only provides a historical perspective on the development of the field but also serves as a foundation for comprehending the complexities of human inheritance. By studying Mendel's experiments, students can gain insights into the basic mechanisms of heredity and appreciate the significance of his contributions to the field of genetics.

Mendelian Inheritance Patterns

In the world of genetics, understanding how traits are passed down from one generation to the next is crucial. This is where Mendelian inheritance patterns come into play. Named after the renowned scientist Gregor Mendel, these patterns help us unravel the secrets behind the transmission of genetic information.

Mendelian inheritance patterns are based on Mendel's groundbreaking experiments with pea plants in the 19th century. He discovered that certain traits, such as flower color or seed shape, are determined by discrete units called genes. These genes exist in pairs, with one copy inherited from each parent. Mendel also observed that some genes are dominant, meaning they mask the effects of other genes, while others are recessive, only exerting their influence when paired with another recessive gene.

One of the simplest Mendelian inheritance patterns is called autosomal dominant inheritance. In this pattern, a single copy of a dominant gene is sufficient to express the trait. For example, if one parent carries the gene for Huntington's disease, there is a 50% chance that their child will inherit the disease. On the other hand, autosomal recessive inheritance requires two copies of the recessive gene for the trait to be expressed. Diseases like cystic fibrosis and sickle cell anemia follow this pattern.

Another Mendelian inheritance pattern is known as X-linked inheritance. This occurs when a gene is located on the X chromosome, one of the sex chromosomes. Since females have two X chromosomes, they can be carriers of X-linked disorders, passing them down to their offspring. Males, however, only have one X chromosome, making them more susceptible to X-linked diseases. Examples include color blindness and hemophilia.

Understanding Mendelian inheritance patterns is essential for genetic counseling, disease diagnosis, and even predicting the likelihood of certain traits in offspring. By studying how genes are inherited, scientists can unravel the complex web of human genetics and provide valuable insights into the causes and treatment of genetic disorders.

In conclusion, Mendelian inheritance patterns provide a framework for understanding how genes are passed down from one generation to the next. By studying these patterns, students of genetics gain valuable insights into the inheritance of traits and the causes of genetic disorders. Whether it is autosomal dominant or recessive inheritance, or the more complex X-linked inheritance, Mendelian patterns are fundamental to our understanding of human genetics. As students, delving into Mendelian inheritance patterns will open a world of knowledge and empower you to make informed decisions in the field of genetics.

Punnett Squares and Genetic Crosses

In the fascinating world of genetics, Punnett squares and genetic crosses are valuable tools used to predict the outcomes of different combinations of genes. Understanding how genes are inherited and the probability of certain traits being passed down is essential to unraveling the mysteries of human genetics. This subchapter aims to provide students with a comprehensive understanding of Punnett squares and genetic crosses, allowing them to explore the intricate world of genetics.

Punnett squares are simple, yet powerful, diagrams that help us visualize the possible combinations of genes that offspring can inherit from their parents. By using these squares, we can determine the likelihood of certain traits appearing in the next generation. This is especially useful when studying traits governed by dominant and recessive alleles.

To begin, let's consider a hypothetical example of a genetic cross between two parents: one with brown eyes (dominant trait) and the other with blue eyes (recessive trait). By representing each parent's alleles for eye color (B for brown and b for blue) on the top and left sides of the Punnett square, we can easily determine the potential eye colors of their offspring. Based on the laws of Mendelian inheritance, we find that the offspring will have a 50% chance of inheriting brown eyes (BB or Bb) and a 50% chance of inheriting blue eyes (bb).

By extending this concept to more complex genetic crosses involving multiple traits, students can unlock the secrets of inheritance patterns and gain a better understanding of genetic diversity. Punnett squares can also be used to explore the inheritance of sex-linked traits, where genes are located on the sex chromosomes.

Furthermore, this subchapter will delve into the significance of genetic crosses in understanding genetic disorders and the principles of inheritance. By examining patterns of inheritance, students can determine the likelihood of inheriting certain genetic conditions, such as cystic fibrosis or sickle cell anemia.

In conclusion, the study of Punnett squares and genetic crosses is crucial for students interested in genetics. This subchapter will equip them with the necessary knowledge and skills to predict the outcome of genetic combinations, understand inheritance patterns, and appreciate the incredible diversity that exists within the human population. Through this exploration, students will gain a deeper understanding of genetics and its impact on human health and diversity.

Pedigree Analysis

Pedigree Analysis: Unraveling the Genetic Story

In the fascinating world of genetics, one powerful tool stands out for understanding the inheritance patterns of genetic traits – pedigree analysis. A pedigree is a visual representation of a family tree that allows us to trace the occurrence of specific traits or genetic disorders across generations. It provides invaluable insights into the inheritance patterns and helps us unlock the mysteries of our genetic makeup. Welcome to the world of pedigree analysis, where we bring DNA to life!

Why is pedigree analysis important? Well, imagine a detective trying to solve a complex case. Pedigree analysis serves as our detective, helping us uncover the patterns and clues hidden within our genes. By examining the family tree, we can identify who has a particular trait, whether it is inherited, and how it is passed down from one generation to the next.

In this chapter, we will delve into the intricacies of pedigree analysis. First, we will learn how to interpret pedigree charts – the language of genetics. Pedigree symbols convey important information about individuals, their gender, and their genetic traits. By understanding these symbols, we can read the story written in our genes.

Next, we will explore the different inheritance patterns that can be observed in pedigrees. From autosomal recessive to X-linked dominant, each pattern offers a unique glimpse into the genetic landscape. We will discuss the characteristics of each pattern and decipher the clues they provide.

Furthermore, we will examine the significance of pedigree analysis in identifying carriers of genetic disorders. By studying pedigrees, we can determine the probability of an individual being a carrier or having a particular disorder. This knowledge is crucial in genetic counseling and making informed decisions about family planning.

Finally, we will investigate how pedigree analysis contributes to the study of human population genetics. By analyzing pedigrees from different populations, we can gain insights into genetic diversity, migration patterns, and evolutionary history.

Pedigree analysis is a powerful tool that unlocks the secrets of our genetic blueprint. It enables us to understand how traits are inherited, predict the probability of genetic disorders, and explore the fascinating story of human evolution. So, let's embark on this journey of discovery, from DNA to diversity, as we uncover the hidden treasures within our genes. Welcome to the world of pedigree analysis!

Chapter 4: Beyond Mendel: Non-Mendelian Inheritance

Incomplete Dominance

In the intricate world of genetics, there are many fascinating concepts that shape the diversity of life. One such concept is incomplete dominance, which unveils a unique pattern of inheritance and adds an extra layer of complexity to our understanding of genetics.

Incomplete dominance occurs when neither allele in a gene pair completely dominates the other. Instead, a blending of the traits from both alleles is observed, resulting in an intermediate phenotype. Imagine a scenario where a red flower and a white flower are crossed. In complete dominance, the offspring would inherit either the red or white color. However, in the case of incomplete dominance, the offspring would display a pink color, a beautiful blend of both parents' traits.

To understand this phenomenon further, let us delve into the underlying genetic mechanisms. At the molecular level, incomplete dominance arises due to the production of an incomplete protein or enzyme by the heterozygous individual carrying two different alleles. This incomplete protein fails to perform its function as efficiently as the complete protein produced by the homozygous individuals.

One classic example of incomplete dominance can be found in the inheritance of flower color in snapdragons. The gene responsible for flower color has two alleles: one for red pigment and another for white pigment. When a snapdragon plant with red flowers is crossed with a snapdragon plant with white flowers, the resulting offspring display pink flowers due to the incomplete dominance between the red and white alleles.

Understanding incomplete dominance is crucial not only for comprehending the complexities of genetics but also for appreciating the diversity found in nature. It challenges the notion of strict dominant and recessive traits, emphasizing the intricate interplay between alleles and the resulting phenotypes.

Moreover, incomplete dominance has important implications in the field of human genetics. Many human traits and diseases exhibit incomplete dominance, such as sickle cell anemia, where individuals with one normal allele and one sickle cell allele have a milder form of the disease. This knowledge is invaluable in the study of inheritance patterns and the development of potential treatments or preventive measures.

In conclusion, incomplete dominance is a captivating concept in genetics that showcases the blending of traits and the emergence of unique phenotypes. By exploring this phenomenon, we gain a deeper understanding of the intricacies of genetics and the remarkable diversity that surrounds us.

Codominance

In the fascinating world of genetics, an important concept that helps us understand the diversity within living organisms is codominance. Codominance refers to a scenario where both alleles of a gene are expressed equally in the phenotype of an individual. This means that instead of one allele dominating or masking the other, both alleles contribute to the observable traits.

To understand codominance, let's consider an example of blood types. In humans, the ABO blood group system is determined by the presence of three alleles: A, B, and O. The A and B alleles are codominant, which means that if an individual inherits both A and B alleles, they will have type AB blood. This is different from other genetic scenarios, such as dominance and recessiveness, where one allele may override the expression of the other.

In codominance, the phenotype reflects a combination or blend of both alleles. For instance, if an individual inherits the A allele from one parent and the B allele from the other parent, their blood type will be AB. This allows for a wide range of possible phenotypes, leading to greater diversity within a population.

Codominance is not limited to blood types; it can be observed in various organisms and traits. In plants, for example, the flower color of certain hybrids may exhibit codominance. If a red-flowered plant is crossed with a white-flowered plant, the resulting hybrid may have flowers that are neither red nor white but rather a combination of both colors, displaying a beautiful blend.

Understanding codominance not only helps us comprehend the complexity of genetic inheritance but also highlights the incredible diversity within species. It demonstrates that organisms possess multiple possibilities when it comes to expressing their traits, leading to the vast array of variations we observe in the natural world.

As students delving into the captivating field of genetics, it is crucial to grasp the concept of codominance. By doing so, we can appreciate the intricate mechanisms behind the diversity of life and recognize the significance of both alleles in shaping an individual's phenotype. So, let's continue our exploration of genetics, discovering the wonders of codominance and unraveling the mysteries of the genetic code that shapes us all.

Multiple Alleles

In the intricate world of genetics, alleles play a crucial role in determining the traits and characteristics that make each one of us unique. You may already be familiar with the concept of alleles – different versions of a gene that can exist at a specific location on a chromosome. However, did you know that genes can have more than just two alleles? Welcome to the fascinating realm of multiple alleles!

Multiple alleles refer to the existence of more than two alternate forms of a gene in a population. While most genes in humans have only two alleles, there are certain genes that have multiple variations. These variations can result in a wide range of phenotypic outcomes, contributing to the diversity we observe in humans.

One of the classic examples of multiple alleles in humans is the ABO blood group system. This system is based on three alleles: A, B, and O. Each individual inherits two alleles, one from each parent, which determine their blood type. The A and B alleles produce specific proteins on the surface of red blood cells, while the O allele does not produce any proteins. The combination of these alleles results in four possible blood types: A, B, AB, and O. This is a prime example of how multiple alleles contribute to the variation in human characteristics.

Another well-known example is the gene responsible for human eye color. While eye color is influenced by several genes, one gene called OCA2 is associated with the production of melanin, the pigment that gives color to the eyes. This gene has multiple alleles that can result in a spectrum of eye colors, ranging from blue to green, hazel, and brown. The presence of different alleles within a population contributes to the diverse eye colors we observe.

Understanding multiple alleles is crucial in comprehending the complexity of genetic inheritance. It highlights the fact that genes are not limited to just two variations but can have a multitude of options. This diversity is what makes each of us unique and contributes to the rich tapestry of human genetics.

Exploring multiple alleles opens up a world of possibilities in the study of genetics. It allows researchers to delve deeper into the complexities of human traits and understand the underlying genetic mechanisms. By studying multiple alleles, we can gain insights into how genetic variation impacts human health, evolution, and the overall diversity of our species.

In conclusion, multiple alleles are a fascinating aspect of genetics that contribute to the incredible diversity we observe in human traits. Whether it is the ABO blood group system or eye color, the presence of multiple alleles adds complexity and richness to our understanding of genetic inheritance. Embracing the concept of multiple alleles opens the door to a deeper exploration of the genetic tapestry that makes us who we are.

Polygenic Inheritance

In the fascinating world of genetics, there are many factors that contribute to the incredible diversity we see in humans. One such factor is polygenic inheritance, which is responsible for many of the complex traits we possess. Polygenic inheritance occurs when multiple genes, each with a small effect, interact to determine a particular trait or characteristic.

Imagine a puzzle, where each piece represents a gene involved in a specific trait. In polygenic inheritance, it takes several puzzle pieces coming together to form the complete picture. Unlike Mendelian inheritance, where a single gene determines a trait, polygenic traits are influenced by the combined effects of multiple genes.

Let's take the example of height. Have you ever wondered why some people are tall while others are short? Well, height is a classic example of a polygenic trait. Multiple genes contribute to our height, each gene adding a small amount to the overall result. This is why you often see a wide range of heights within a family or a population.

Polygenic traits are not just limited to physical characteristics like height. They also play a role in determining traits such as skin color, eye color, and even susceptibility to certain diseases. These traits are influenced by the interaction of many genes, each with their own contribution.

The study of polygenic inheritance has revealed some fascinating findings. For example, researchers have discovered that certain combinations of genes can increase the likelihood of developing diseases like diabetes, heart disease, or even certain types of cancer.

Understanding the complex interactions between genes can help us predict and potentially prevent the onset of these diseases.

In conclusion, polygenic inheritance is a fascinating aspect of human genetics that contributes to the incredible diversity we see in our population. It explains why traits like height, skin color, and susceptibility to diseases vary from person to person. By studying polygenic inheritance, scientists can gain a deeper understanding of how our genes interact and potentially develop new strategies for preventing and treating diseases. As students of genetics, exploring the realm of polygenic inheritance opens up a world of possibilities for further research and discovery.

Pleiotropy

Pleiotropy: Unraveling the Complexity of Genetics

Genetics is a captivating field that explores the intricate mechanisms behind the inheritance and expression of traits. Within this astonishing realm, a phenomenon called pleiotropy arises, adding another layer of complexity to the genetic landscape. Pleiotropy refers to the ability of a single gene to influence multiple, seemingly unrelated traits. This subchapter will delve into the fascinating world of pleiotropy, unraveling its significance and implications in the realm of human genetics.

Imagine a gene as a master puppeteer, pulling strings that govern various aspects of our biological makeup. Pleiotropy occurs when this puppeteer gene not only controls one trait but orchestrates a multitude of characteristics. It is akin to a single conductor guiding an entire symphony orchestra. For instance, a gene responsible for pigmentation might also contribute to the development of eye color, hair texture, and even certain diseases or disorders.

Understanding pleiotropy is crucial because it challenges the traditional Mendelian concept of a one-to-one relationship between genes and traits. Instead, pleiotropy demonstrates the intricate interplay between genes, revealing the complexity underlying human genetic variation. By studying pleiotropic genes, scientists gain insights into the shared molecular pathways that regulate multiple traits, potentially paving the way for breakthroughs in personalized medicine.

Moreover, pleiotropy sheds light on the concept of genetic trade-offs. As genes influence several traits simultaneously, changes in one characteristic may have unintended consequences on others. This phenomenon can be observed in the persistence of certain genetic disorders, where a beneficial effect on one trait may come at the cost of increased susceptibility to another condition.

Pleiotropy also plays a significant role in evolutionary biology. Genes that exhibit pleiotropic effects can drive the diversification and adaptation of species over time. For example, a gene responsible for the elongation of bird beaks may also influence the shape and size of their wings, contributing to their survival and success in different environments.

In conclusion, pleiotropy is a captivating aspect of genetics that unravels the intricacies of gene expression and inheritance. By understanding pleiotropic genes, students of genetics gain a deeper appreciation for the complexity of the human genome and its role in shaping our traits and health. From personalized medicine to evolutionary biology, pleiotropy offers a window into the interconnectedness of genetic variation and the remarkable diversity of life forms on our planet.

Chapter 5: Human Genetic Disorders

Autosomal Dominant Disorders

In the vast realm of human genetics, there are various types of genetic disorders that can affect individuals. One such category is known as autosomal dominant disorders. These disorders are caused by mutations in genes located on autosomal chromosomes, which are non-sex chromosomes. Unlike autosomal recessive disorders, where two copies of the mutated gene are required to manifest the disorder, autosomal dominant disorders only require one copy of the mutated gene for the disorder to be expressed.

Autosomal dominant disorders are characterized by their inheritance pattern, which means they can be passed down from one generation to the next. If a parent has the disorder, there is a 50% chance that their child will inherit the mutated gene and develop the disorder as well. This means that even if only one parent carries the gene, their child has a chance of being affected.

Some well-known autosomal dominant disorders include Huntington's disease, Marfan syndrome, and neurofibromatosis. Each of these disorders has its own unique set of symptoms and complications. For example, Huntington's disease is a progressive neurological disorder that affects movement, cognition, and behavior. Marfan syndrome, on the other hand, affects the connective tissues in the body and can lead to heart problems, skeletal abnormalities, and eye complications. Neurofibromatosis is characterized by the growth of tumors on nerves and can cause various symptoms depending on the location of the tumors.

Understanding the mechanisms behind autosomal dominant disorders is crucial in the field of genetics. Researchers and scientists work tirelessly to identify the specific genes responsible for these disorders and uncover the underlying molecular processes. This knowledge not only aids in the development of diagnostic tests but also plays a vital role in potential treatments and therapies.

For students interested in genetics, studying autosomal dominant disorders can provide valuable insights into the complexities of human genetics. Exploring these disorders can help students understand how genetic mutations can affect an individual's health and contribute to the overall diversity within the human population. Additionally, understanding the inheritance patterns of these disorders can shed light on the importance of genetic counseling and family planning for individuals and families at risk of passing on these disorders.

In conclusion, autosomal dominant disorders are a fascinating subset of genetic disorders that have significant implications for human health. By delving into the intricacies of these disorders, students can gain a deeper understanding of genetics and its impact on our lives.

Autosomal Recessive Disorders

In the realm of human genetics, there are various types of genetic disorders that can affect individuals. One significant category is known as autosomal recessive disorders. These disorders occur when an individual inherits two copies of a mutated gene, one from each parent, resulting in a malfunctioning or absent protein.

Autosomal recessive disorders differ from autosomal dominant disorders, as the latter only require one copy of the mutated gene to manifest the disorder. In the case of recessive disorders, the unaffected parent is often referred to as a carrier, as they possess one copy of the mutated gene but do not exhibit any symptoms themselves.

One example of an autosomal recessive disorder is cystic fibrosis (CF), a disease that primarily affects the lungs and digestive system. CF occurs due to a mutation in the CFTR gene, which leads to the production of a faulty protein. As a result, individuals with CF experience thick mucus buildup in their lungs and other organs, leading to breathing difficulties and digestive problems.

Another autosomal recessive disorder is sickle cell disease (SCD), a condition that affects red blood cells. SCD is caused by a mutation in the HBB gene, which alters the structure of hemoglobin, the protein responsible for carrying oxygen in the blood. This mutation causes red blood cells to become misshapen, leading to chronic pain, anemia, and other complications.

It is important for students studying genetics to understand the inheritance patterns of autosomal recessive disorders. In order for an individual to develop an autosomal recessive disorder, both parents must be carriers of the mutated gene. When these carriers have children, there is a 25% chance that the child will inherit two copies of the mutated gene and develop the disorder, a 50% chance of being a carrier like their parents, and a 25% chance of not inheriting the mutated gene at all.

Genetic counseling plays a crucial role in the management and prevention of autosomal recessive disorders. By identifying carriers within families, individuals can make informed decisions about family planning, such as undergoing prenatal testing or considering assisted reproductive technologies.

Autosomal recessive disorders serve as a reminder of the complexity and diversity of human genetics. Understanding these disorders is essential for students studying genetics, as it provides insights into the inheritance patterns and the impact of genetic mutations on human health. Through continued research and education, we strive to uncover new treatments and preventive measures to improve the lives of individuals affected by autosomal recessive disorders.

X-Linked Disorders

In the world of genetics, there are various types of genetic disorders that can affect individuals in unique ways. One such category is X-linked disorders. These disorders are associated with genes located on the X chromosome, one of the two sex chromosomes.

To understand X-linked disorders, it is important to first grasp the basics of sex determination in humans. Females have two X chromosomes (XX), while males have one X and one Y chromosome (XY). Since males have only one X chromosome, any genetic mutation on that chromosome can have a significant impact on their health.

X-linked disorders are typically recessive, meaning that a male with a single copy of the mutated gene on his X chromosome will exhibit the disorder. Females, on the other hand, need to have two copies of the mutated gene, one on each X chromosome, to manifest the disorder. This is because the presence of a normal gene on one of the X chromosomes can compensate for the mutation.

Examples of X-linked disorders include hemophilia, Duchenne muscular dystrophy, and color blindness. Hemophilia is a bleeding disorder characterized by the inability of blood to clot properly, while Duchenne muscular dystrophy is a progressive muscle-weakening condition. Color blindness affects an individual's ability to perceive certain colors accurately.

While X-linked disorders predominantly affect males, females can also be carriers of these conditions. A carrier is an individual who has one copy of the mutated gene but does not exhibit symptoms. Carriers can pass on the mutated gene to their offspring, potentially affecting future generations.

Understanding X-linked disorders is crucial for geneticists and medical professionals working in the field of genetics. By studying these disorders, researchers can gain insights into the underlying mechanisms and develop potential treatments or therapies.

For students interested in genetics, learning about X-linked disorders provides a fascinating glimpse into the complexity of human genetics. By understanding the inheritance patterns and the impact of these disorders, students can appreciate the importance of genetic counseling and the potential implications for individuals and families affected by X-linked disorders.

Overall, X-linked disorders represent an intriguing aspect of human genetics. Exploring these disorders can deepen our understanding of how genetics influences our health and well-being, and pave the way for advancements in diagnosis, treatment, and prevention of genetic diseases.

Genetic Counseling and Testing

Genetic counseling and testing play a crucial role in the field of genetics, helping individuals and families make informed decisions regarding their health and future. This subchapter aims to provide students with an introduction to genetic counseling and testing, exploring the significance, process, and ethical considerations surrounding these practices.

Genetic counseling is a process that involves an individual or couple meeting with a trained professional to discuss their genetic makeup, inherited conditions, and potential risks associated with certain genetic disorders. Genetic counselors are highly skilled in interpreting complex genetic information and assisting individuals in understanding the implications of their genetic test results.

One of the primary goals of genetic counseling is to empower individuals to make informed decisions about their reproductive choices, medical management, and lifestyle. Genetic counselors provide support, guidance, and information to help individuals understand their genetic risk factors, genetic testing options, and available preventive measures or treatments.

Genetic testing is a laboratory process that can reveal changes or mutations in an individual's genes or chromosomes. There are different types of genetic tests, such as diagnostic, carrier, predictive, and prenatal testing. Diagnostic testing helps identify the cause of a known genetic condition or diagnose a new one. Carrier testing determines if an individual carries a gene mutation that could be passed on to their children. Predictive testing assesses the likelihood of developing a genetic condition later in life. Prenatal testing is performed during pregnancy to detect genetic disorders in the fetus.

While genetic testing provides valuable information, it also raises ethical concerns. The subchapter discusses the importance of informed consent in genetic testing, ensuring individuals have a clear understanding of the risks, benefits, and limitations of the tests. It also explores the potential psychological and social implications of genetic testing, as it can uncover unexpected results or lead to difficult decisions for individuals and families.

Additionally, the subchapter delves into the ethical considerations surrounding genetic counseling and testing, such as privacy, confidentiality, and discrimination based on genetic information. It highlights the importance of protecting individuals' genetic information and promoting fair treatment and equal opportunities for all.

In conclusion, genetic counseling and testing are essential tools in the field of genetics, enabling individuals to make informed decisions about their health and reproductive choices. Understanding the significance, process, and ethical considerations of these practices is crucial for students interested in the field of genetics, as they play a vital role in shaping the future of healthcare and the well-being of individuals and families.

Ethical Considerations in Genetic Testing

In the rapidly advancing field of genetics, the development of genetic testing has opened up new possibilities for understanding and predicting human health. Genetic testing involves analyzing a person's DNA to identify potential genetic disorders or to determine their risk for certain diseases. While this technology has the potential to revolutionize healthcare, it also raises important ethical considerations that must be carefully examined.

One of the primary ethical considerations in genetic testing is the issue of informed consent. Students studying genetics need to understand that genetic testing should only be conducted with the explicit consent of the individual being tested. This means that the person must fully understand the purpose, risks, and potential outcomes of the test before giving their consent. Informed consent is crucial to ensuring that individuals maintain their autonomy and have control over their genetic information.

Another ethical concern is the potential for discrimination based on genetic test results. Genetic information is highly personal and sensitive, and there is a risk that this information could be used against individuals by employers, insurers, or even potential partners. Students need to be aware of the legal protections in place, such as the Genetic Information Nondiscrimination Act (GINA), which prohibits discrimination based on genetic information in employment and health insurance. It is important for students to advocate for the protection of genetic privacy and to understand the potential consequences of sharing their genetic information.

Genetic testing also raises questions about the appropriate use of the information obtained. For example, should parents have access to genetic information about their unborn child? Should individuals be able to access their own genetic test results without the involvement of a healthcare professional? These are complex ethical dilemmas that require careful consideration. Students need to be aware of the potential benefits and risks associated with different approaches to the use and dissemination of genetic information.

Finally, the issue of genetic testing in vulnerable populations must be addressed. There is a risk that certain groups may be disproportionately affected by the ethical implications of genetic testing, such as minority populations or individuals with disabilities. Students need to be mindful of the potential for disparities in access to and interpretation of genetic test results, and they should advocate for equitable access to genetic testing and counseling services.

In conclusion, genetic testing holds great promise for improving human health, but it also raises important ethical considerations. Students studying genetics must be aware of the need for informed consent, the risk of discrimination, the appropriate use of genetic information, and the potential disparities in vulnerable populations. By understanding and addressing these ethical considerations, students can contribute to the responsible and ethical use of genetic testing in the field of genetics.

Chapter 6: Chromosomal Abnormalities

Types of Chromosomal Abnormalities

In the world of genetics, chromosomal abnormalities play a significant role in shaping the diversity of human beings. The human body consists of trillions of cells, each containing 46 chromosomes, which are thread-like structures made up of DNA molecules. However, sometimes errors occur during the process of cell division, leading to changes in the structure or number of chromosomes. These changes are known as chromosomal abnormalities and can have profound effects on an individual's health and development.

There are several types of chromosomal abnormalities that students of genetics should be aware of. One of the most common types is called Down syndrome, also known as trisomy 21. In this condition, individuals have an extra copy of chromosome 21, resulting in intellectual disabilities, distinct facial features, and potential health issues such as heart defects. Down syndrome occurs in approximately 1 in every 700 births and is often associated with advanced maternal age.

Another type of chromosomal abnormality is Turner syndrome, which affects only females. It occurs when one of the two X chromosomes is either missing or partially missing. Women with Turner syndrome often have short stature, infertility, and certain physical features such as a webbed neck and low-set ears. It is estimated that 1 in every 2,500 female births is affected by Turner syndrome.

Klinefelter syndrome is a chromosomal abnormality that affects males. It occurs when a male is born with an extra X chromosome, resulting in a total of 47 chromosomes instead of the usual 46. Men with Klinefelter syndrome may experience infertility, reduced muscle mass, enlarged breasts, and learning difficulties. Approximately 1 in every 500 males is affected by Klinefelter syndrome.

Other chromosomal abnormalities include cri du chat syndrome, caused by a deletion on chromosome 5, and Patau syndrome, caused by an extra copy of chromosome 13. Each of these conditions presents unique challenges and can impact an individual's physical and intellectual capabilities.

Understanding the various types of chromosomal abnormalities is crucial for students of genetics. By learning about these conditions, students can gain insight into the complexity of human genetics and the factors that contribute to human diversity. Additionally, this knowledge can help students better comprehend the impact of chromosomal abnormalities on individuals and society as a whole.

Down Syndrome

Down Syndrome is a genetic disorder that affects individuals from birth. It is caused by the presence of an extra copy of chromosome 21, resulting in a total of three copies instead of the usual two. This additional genetic material disrupts the normal development and functioning of the body and brain.

One of the most noticeable characteristics of individuals with Down Syndrome is their distinct facial features, such as almond-shaped eyes, a flat nasal bridge, and a protruding tongue. They also tend to have lower muscle tone and shorter stature compared to their peers. However, it is important to remember that every person with Down Syndrome is unique and may exhibit varying degrees of these physical traits.

Apart from physical characteristics, individuals with Down Syndrome may also experience certain health issues. They are more prone to congenital heart defects, hearing and vision problems, and respiratory infections. However, with early medical intervention and appropriate care, many of these conditions can be managed effectively.

Cognitive abilities can also be affected in individuals with Down Syndrome. Most people with this condition have mild to moderate intellectual disability, but their cognitive abilities can vary widely. It is important to provide support and tailored education to help individuals with Down Syndrome reach their full potential.

Despite the challenges associated with Down Syndrome, individuals with this condition can lead fulfilling and meaningful lives. With the right support from family, friends, and the community, they can participate in school, work, and recreational activities, just like their peers.

In recent years, there have been significant advancements in medical and educational interventions for individuals with Down Syndrome. Early intervention programs, speech therapy, physical therapy, and special education programs have greatly improved the quality of life for those affected by this condition.

It is crucial to foster understanding and acceptance of individuals with Down Syndrome within our society. By educating ourselves about this genetic disorder and promoting inclusivity, we can create a more compassionate and inclusive world for all.

In conclusion, Down Syndrome is a genetic disorder caused by the presence of an extra copy of chromosome 21. It affects physical characteristics, cognitive abilities, and overall development. With early intervention and support, individuals with Down Syndrome can lead fulfilling lives and contribute to our diverse society.

Turner Syndrome

Turner Syndrome is a genetic disorder that affects only females. It occurs when one of the two X chromosomes is either completely or partially missing. This condition is named after Dr. Henry Turner, who first described it in 1938. It is estimated that about 1 in every 2,500 girls is born with Turner Syndrome.

Girls with Turner Syndrome may exhibit a range of physical and developmental characteristics. One of the most common signs is short stature. These girls tend to be shorter than their peers, and their growth may be slower during childhood and puberty. Additionally, they may have a webbed neck, low-set ears, and a broad chest. Some girls with Turner Syndrome also have certain heart and kidney abnormalities.

Turner Syndrome can also affect a girl's reproductive system. Most girls with this condition are infertile, meaning they cannot conceive children naturally. They usually do not undergo puberty without treatment, and hormone therapy is often necessary to induce the development of secondary sexual characteristics, such as breast development and menstruation.

Aside from physical traits, Turner Syndrome can also have an impact on a girl's cognitive and social development. Many girls with Turner Syndrome experience learning difficulties, particularly in math and spatial skills. They may also have difficulties with social interactions and may benefit from specialized support and education.

Diagnosis of Turner Syndrome typically occurs during childhood or adolescence. Doctors may suspect the condition based on physical characteristics and growth patterns. A blood test called a karyotype analysis is used to confirm the diagnosis by examining the chromosomes. Early detection is essential to provide appropriate medical and psychological support to girls with Turner Syndrome.

While there is no cure for Turner Syndrome, ongoing medical care and support can help manage the symptoms and improve quality of life. Hormone therapy can help girls reach a normal height and develop secondary sexual characteristics. Regular check-ups are necessary to monitor heart and kidney health, as well as to address any educational or social challenges that may arise.

In conclusion, Turner Syndrome is a genetic disorder that affects females and is characterized by the absence or partial absence of one X chromosome. It can impact physical, reproductive, cognitive, and social development. Early diagnosis and appropriate medical care are crucial for managing the condition and providing necessary support to affected individuals.

Klinefelter Syndrome

Klinefelter Syndrome is a genetic disorder that affects males. It occurs when a male is born with an extra X chromosome, resulting in a total of 47 chromosomes instead of the usual 46. This extra chromosome is usually inherited from the mother.

The symptoms of Klinefelter Syndrome can vary from person to person, but some common signs include taller stature, reduced muscle mass, smaller testes, and infertility. Additionally, individuals with this condition may have learning disabilities, language difficulties, and problems with social skills. However, it is important to note that not all individuals with Klinefelter Syndrome experience these symptoms, and the severity can vary widely.

The diagnosis of Klinefelter Syndrome is typically made through genetic testing, such as a karyotype analysis. This test examines a person's chromosomes to determine if there are any abnormalities. Early diagnosis is crucial as it allows for appropriate medical management and support.

While there is no cure for Klinefelter Syndrome, there are various treatments and interventions that can help manage the symptoms. Hormone replacement therapy (HRT) is commonly used to address hormonal imbalances and promote the development of secondary sexual characteristics. Speech and occupational therapy may be beneficial for individuals experiencing speech and motor skill difficulties. In some cases, fertility treatments can also be considered to help individuals with Klinefelter Syndrome have children.

Living with Klinefelter Syndrome can present challenges, but with the right support and interventions, individuals can lead fulfilling lives. It is essential for individuals with Klinefelter Syndrome to have a supportive network of healthcare professionals, educators, and family members who can provide guidance and understanding.

Research into Klinefelter Syndrome is ongoing, aiming to better understand the condition and develop new treatments. Scientists are exploring the underlying genetic mechanisms that lead to the symptoms of Klinefelter Syndrome, which may pave the way for more targeted therapies in the future.

By raising awareness about Klinefelter Syndrome, we can help promote understanding and acceptance for individuals with this genetic disorder. Through education and support, we can ensure that every individual affected by Klinefelter Syndrome has the opportunity to thrive and reach their full potential.

Genetic Disorders Resulting from Chromosomal Abnormalities

Understanding the role of genetics in human health is essential for students studying genetics. Genetic disorders can occur due to a variety of factors, one of which is chromosomal abnormalities. In this subchapter, we will explore the impact of chromosomal abnormalities on human health and the genetic disorders that can result from these abnormalities.

Chromosomes are thread-like structures composed of DNA that carry our genetic information. Humans typically have 46 chromosomes, with 23 inherited from each parent. However, chromosomal abnormalities can occur when there is an error in the number or structure of chromosomes during cell division.

One common chromosomal abnormality is Down syndrome, also known as trisomy 21. Individuals with Down syndrome have an extra copy of chromosome 21, resulting in intellectual disabilities, characteristic facial features, and an increased risk for certain health conditions. Students will learn about the genetic basis of Down syndrome and the impact it has on individuals and their families.

Another example of a genetic disorder resulting from chromosomal abnormalities is Turner syndrome. This disorder occurs in females who are missing or have incomplete X chromosomes. Students will explore the features of Turner syndrome, including short stature, infertility, and potential heart and kidney problems. They will also learn about the genetic mechanisms that lead to this condition.

Students will also be introduced to other chromosomal disorders such as Klinefelter syndrome, XYY syndrome, and fragile X syndrome.

These disorders result from additional or missing sex chromosomes and can have varying effects on individuals' physical and intellectual development. Understanding the genetic basis of these disorders will provide students with a comprehensive understanding of the impact of chromosomal abnormalities on human health.

Furthermore, this subchapter will delve into the genetic counseling and testing options available for individuals and families at risk of chromosomal abnormalities. Students will learn about prenatal testing methods, such as chorionic villus sampling and amniocentesis, and how these tests can help identify chromosomal disorders early on. Genetic counseling, which involves providing information and support to families, will also be discussed as an important resource for individuals affected by chromosomal abnormalities.

By exploring genetic disorders resulting from chromosomal abnormalities, students will gain a deeper understanding of the intricate relationship between genetics and human health. This knowledge will enable them to make informed decisions and contribute to the field of genetics in the future.

Chapter 7: Genetic Variation and Evolution

Genetic Variation in Populations

In the fascinating world of genetics, one of the most crucial concepts to understand is genetic variation in populations. Genetic variation refers to the diversity of genes within a particular population. It is this variation that gives rise to the incredible diversity we see in the world around us, from the different physical traits we possess to the wide range of diseases that affect individuals.

Understanding genetic variation is essential to grasp how populations evolve and adapt over time. It is the driving force behind natural selection, where certain traits become more or less common in a population due to their survival and reproductive advantages. Without genetic variation, species would lack the ability to adapt to changing environments, making them vulnerable to extinction.

There are several sources of genetic variation in populations. One of the primary sources is mutations, which are spontaneous changes in the DNA sequence. These mutations can occur during DNA replication or can be caused by external factors such as radiation or chemicals. While most mutations are neutral or harmful, occasionally, they can introduce a beneficial change that provides an advantage to individuals carrying that particular variation.

Another significant source of genetic variation is genetic recombination, which occurs during the formation of reproductive cells. This process, known as meiosis, shuffles the genetic material from an individual's parents, creating unique combinations of genes in each offspring. Genetic recombination not only increases the genetic diversity within a population but also ensures that no two individuals are genetically identical.

The study of genetic variation has numerous practical applications. It allows scientists to trace the ancestry of individuals and populations, providing insights into human migration patterns and the evolutionary history of our species. Additionally, understanding genetic variation is crucial in the field of medicine, as it helps identify genetic factors that contribute to diseases and develop personalized treatments.

As students of genetics, it is essential to comprehend the significance of genetic variation in populations. It is through this variation that life on Earth has evolved and continues to adapt to ever-changing environments. By studying and appreciating genetic diversity, we gain a deeper understanding of the intricate mechanisms that shape life as we know it.

Natural Selection and Adaptation

Subchapter: Natural Selection and Adaptation

Introduction:

In the vast tapestry of life, the process of natural selection has shaped the incredible diversity of species that inhabit our planet. From the tiniest microorganisms to the majestic creatures roaming the Earth, the principles of genetics and evolution have played a crucial role in sculpting the complex web of life. In this subchapter, we will explore the fascinating concepts of natural selection and adaptation and delve into the intricate mechanisms that drive genetic diversity.

Understanding Natural Selection:

Natural selection can be thought of as nature's way of selecting the fittest individuals within a population to survive and reproduce, resulting in the propagation of advantageous traits over time. In the struggle for existence, individuals with traits that confer a survival advantage are more likely to pass on their genetic material to the next generation. This process of differential reproductive success drives evolutionary change and ultimately leads to the adaptation of species to their environments.

The Role of Genetics:

Genetics lies at the heart of natural selection. Variations in genes, which are inherited from our parents, give rise to the traits we possess. Some of these traits may increase an organism's chances of survival and reproduction, while others may be less advantageous. Through the interplay of genetic variation and the selective pressures exerted by the environment, populations evolve over time, leading to the emergence of new species and the diversification of life.

Adaptive Traits:

Adaptation is the process by which a population becomes better suited to its environment over generations. Certain traits, such as camouflage, mimicry, or the ability to withstand extreme temperatures, can enhance an organism's chances of survival and reproduction. These adaptive traits arise through natural selection, as individuals with these traits are more likely to survive and pass on their genes.

Examples of Natural Selection:

Numerous examples of natural selection can be observed in the natural world. From the long neck of a giraffe, which allows it to reach the leaves of tall trees, to the antibiotic resistance of bacteria, which evolves in response to the overuse of antibiotics, these adaptations highlight the power of natural selection to shape the diversity of life.

Conclusion:

The study of natural selection and adaptation is a cornerstone of genetics, providing insights into the incredible diversity and complexity of the natural world. By understanding the mechanisms that drive evolution, we gain a deeper appreciation for the interconnectedness of all living things and the remarkable power of nature's own genetic experiments. As students of genetics, let us embark on this journey of discovery, unraveling the secrets of life's tapestry and unlocking the mysteries of our own genetic makeup.

Genetic Drift and Gene Flow

In the intriguing world of genetics, two fundamental processes play a significant role in shaping the diversity of species: genetic drift and gene flow. These processes, while distinct, can both have profound effects on the genetic makeup of populations.

Genetic drift refers to the random fluctuations in the frequency of genetic traits within a population over time. This phenomenon is particularly influential in small populations where chance events can have a substantial impact. Imagine a small island with a limited number of individuals. If, by chance, a particular genetic trait becomes more prevalent in the first generation, its frequency may increase even further through subsequent generations due to the limited genetic variation. This random fluctuation in gene frequency is known as the founder effect. Over time, genetic drift can lead to the fixation or loss of specific traits within a population, contributing to the overall genetic diversity or homogeneity.

On the other hand, gene flow is the movement of genetic material between populations through migration or interbreeding. When individuals from separate populations mate and exchange genes, they introduce new genetic variation into the recipient population. This process helps to prevent genetic differentiation and maintain genetic diversity within a species. For instance, imagine a population of birds on one side of a mountain range and another population on the opposite side. If a few individuals from one population manage to cross the mountains and mate with individuals from the other side, they will introduce new genetic material to the previously isolated population, reducing genetic differences between the two groups.

Understanding genetic drift and gene flow is crucial in the study of human genetics. These processes play a vital role in shaping the genetic diversity within and between populations, influencing the occurrence of genetic disorders and the response to environmental factors. By comprehending the mechanisms of genetic drift and gene flow, we can better appreciate how genetic variation arises and how it impacts the human population as a whole.

In conclusion, genetic drift and gene flow are two essential processes that influence the diversity and genetic makeup of populations. Genetic drift is characterized by random fluctuations in gene frequency, while gene flow refers to the movement of genetic material between populations. Both processes contribute to the overall genetic diversity and are central to the study of human genetics. By examining these mechanisms, we can gain a deeper understanding of how genetic variation arises and its implications for human populations.

Human Evolution and Migration

Introduction:
In the fascinating field of genetics, the study of human evolution and migration holds great significance. Understanding our ancient past and the journeys our ancestors undertook helps shed light on our present genetic diversity. In this subchapter, we will delve into the captivating story of human evolution and migration, exploring how our species spread across the globe and adapted to various environments.

Human Origins:
Our journey begins in Africa, where modern humans, Homo sapiens, originated around 200,000 years ago. We share a common ancestry with other hominin species such as Neanderthals and Denisovans, who lived alongside our early ancestors. By analyzing ancient DNA, scientists have been able to unravel the complex web of relationships between these species and gain insights into our evolutionary history.

Out of Africa:
Around 70,000 years ago, a pivotal event occurred in human history – the Out of Africa migration. A small group of Homo sapiens left Africa, embarking on a journey that would eventually lead to our global presence. These intrepid explorers migrated to different parts of the world, encountering diverse environments and adapting to new challenges along the way.

Genetic Diversity:
As humans dispersed across the globe, genetic diversity began to emerge. Over time, isolated populations evolved distinct genetic traits,

influenced by factors such as natural selection, genetic drift, and cultural practices. By studying the genetic makeup of different populations, scientists can trace the migration routes of our ancestors and uncover the genetic footprints they left behind.

Adaptation and Natural Selection:
Migration not only shaped our genetic diversity but also facilitated adaptation to different environments. As humans settled in new regions, they faced unique challenges, such as climate extremes and different diets. Natural selection favored genetic variations that allowed individuals to survive and reproduce in these environments. The study of genetic adaptations provides valuable insights into how our ancestors thrived in diverse landscapes.

Modern Human Diversity:
Today, humanity exhibits a remarkable range of physical and genetic diversity. This diversity is a testament to the complex history of migration and adaptation that our species has undergone. By understanding the genetic differences among populations, we can gain insights into the origins of various traits and diseases, as well as the interconnectedness of all humans.

Conclusion:
The study of human evolution and migration is a captivating journey through time. Through the lens of genetics, we can unravel the story of our species, tracing our roots back to Africa and exploring the incredible diversity that exists among us. By understanding our past, we can better appreciate the interconnectedness of all humans and the shared journey we have embarked upon.

Chapter 8: Applications of Human Genetics

Pharmacogenetics

In the rapidly advancing field of genetics, one branch that holds great promise for the future of medicine is pharmacogenetics. Pharmacogenetics explores how an individual's genetic makeup influences their response to drugs, both in terms of effectiveness and potential side effects. This subchapter will delve into the fascinating world of pharmacogenetics, providing students with an understanding of how genetics can play a crucial role in personalized medicine.

Pharmacogenetics is founded on the principle that each person's genetic composition is unique, affecting their ability to metabolize drugs and respond to treatment. By studying the genetic variations that influence drug response, scientists can identify individuals who are likely to benefit from certain medications, as well as those who may experience adverse reactions. This knowledge can help healthcare professionals make informed decisions about drug dosage and selection, improving patient outcomes and reducing the risk of adverse events.

Students will learn about the different types of genetic variations that can impact drug response, such as single nucleotide polymorphisms (SNPs) and copy number variations (CNVs). Through engaging explanations and real-life examples, they will gain an understanding of how these genetic variations can alter drug metabolism enzymes, drug targets, and transporters, ultimately affecting drug efficacy and toxicity.

Furthermore, this subchapter will cover the role of pharmacogenetics in various medical specialties, such as oncology, psychiatry, and cardiology. Students will explore how pharmacogenetic testing can guide medication choices for cancer patients, ensuring personalized treatment plans that optimize therapeutic outcomes. They will also discover how pharmacogenetics can help predict an individual's response to psychiatric medications, leading to more effective and tailored mental health treatments.

To make the content even more engaging and relatable, case studies of real patients who have benefited from pharmacogenetic testing will be included. These stories will emphasize the impact of personalized medicine and highlight the potential of pharmacogenetics to revolutionize healthcare.

By the end of this subchapter, students will have a solid understanding of the principles of pharmacogenetics and its potential applications in various medical fields. They will appreciate the importance of genetic variation in drug response and gain insight into the future of personalized medicine. With this knowledge, they will be equipped to explore the exciting world of genetics and its implications for improving patient care.

Personalized Medicine

In recent years, the field of genetics has revolutionized the way we think about medicine. No longer is healthcare a one-size-fits-all approach, but rather a personalized and tailored experience. This subchapter will explore the concept of personalized medicine and its implications in the field of genetics.

Personalized medicine is an emerging field that takes into account an individual's unique genetic makeup, lifestyle, and environment to determine the best course of treatment. It recognizes that each person is different and that one treatment may not be effective for everyone. By analyzing an individual's genetic information, doctors can gain valuable insights into their predisposition for certain diseases, how their body metabolizes medications, and how they may respond to various treatments.

Advancements in genetic technologies, such as whole genome sequencing, have made it possible to gather vast amounts of genetic data quickly and affordably. This wealth of information allows researchers and healthcare professionals to identify genetic variations that may contribute to an individual's risk of developing certain diseases. Armed with this knowledge, doctors can develop personalized prevention strategies and treatment plans that target the root cause of the disease.

For example, let's consider a hypothetical student named Jane. Jane's genetic analysis reveals that she has a high risk of developing heart disease due to a specific gene variant. Armed with this information, Jane's doctor may recommend lifestyle changes such as a heart-healthy diet and regular exercise, as well as prescribe medications that specifically target her genetic predisposition to heart disease. By tailoring her treatment to her genetic profile, Jane has a higher chance of preventing or effectively managing her condition.

Personalized medicine also holds great promise in the field of pharmacogenomics. This branch of genetics focuses on how an individual's genetic makeup influences their response to medications. By analyzing a person's genetic variants, doctors can predict how they will metabolize certain drugs, allowing for more precise dosing and minimizing potential side effects.

As students interested in genetics, understanding the concept of personalized medicine is crucial. This field is at the forefront of medical advancements, and it has the potential to revolutionize the way we approach healthcare. By recognizing the diversity of our genetic makeup, healthcare professionals can provide more effective treatments that are tailored to each individual's needs. As future scientists, researchers, and healthcare professionals, you have the opportunity to contribute to the advancements in personalized medicine and shape the future of healthcare.

Forensic Genetics

Forensic genetics is a fascinating field of study that blends the science of genetics with the investigation of crimes. It involves the analysis of DNA samples found at crime scenes to identify and link suspects to the scene, as well as to establish relationships between individuals. This subchapter will introduce you, as students interested in genetics, to the exciting world of forensic genetics.

DNA, or deoxyribonucleic acid, is the molecule that carries the genetic instructions for the development and functioning of all living organisms. It is present in every cell of our bodies and is unique to each individual, except for identical twins. Forensic geneticists utilize this uniqueness to identify individuals and provide crucial evidence in criminal investigations.

One of the primary techniques used in forensic genetics is DNA profiling, also known as DNA fingerprinting. This technique compares specific regions of an individual's DNA to determine if it matches DNA found at a crime scene. The process involves extracting DNA from cells, amplifying specific regions using a technique called polymerase chain reaction (PCR), and then analyzing the resulting DNA fragments using a method called gel electrophoresis. DNA profiling has revolutionized criminal investigations by helping to solve cold cases, exonerate innocent individuals, and ensure justice is served.

Another important application of forensic genetics is determining paternity or establishing family relationships. By comparing DNA profiles of individuals, geneticists can determine if two individuals are related and, if so, the degree of relatedness. This information has been instrumental in resolving legal disputes, such as child custody cases or inheritance claims.

Moreover, advancements in forensic genetics have led to the development of specialized databases, such as the Combined DNA Index System (CODIS), which stores DNA profiles of known criminals and unidentified crime scene samples. These databases enable investigators to match DNA profiles from crime scenes to known criminals, helping to solve cases and prevent future crimes.

As students interested in genetics, learning about forensic genetics provides a unique perspective on how genetic knowledge can be applied in real-world scenarios. It demonstrates the power and impact of genetics beyond the traditional fields of medicine and research.

In conclusion, forensic genetics plays a crucial role in criminal investigations by utilizing the unique properties of DNA to identify individuals and establish relationships. The techniques and databases developed in this field have revolutionized the way crimes are solved and justice is served. Understanding forensic genetics not only broadens our knowledge of genetics but also highlights the practical applications of genetic science in society.

Genetic Engineering and Biotechnology

In the ever-evolving field of genetics, one of the most exciting and controversial areas of study is genetic engineering and biotechnology. These groundbreaking technologies have the potential to revolutionize medicine, agriculture, and our understanding of the natural world. In this subchapter, we will explore the basics of genetic engineering and biotechnology, and how they are shaping our world today.

Genetic engineering involves manipulating an organism's DNA to alter its characteristics or introduce new traits. This process is accomplished through various techniques, such as gene editing using CRISPR-Cas9, which allows scientists to precisely cut and modify specific genes. Genetic engineering has opened up possibilities for curing genetic diseases, increasing crop yields, and even creating genetically modified organisms (GMOs) that are resistant to pests or able to thrive in harsh environments.

Biotechnology, on the other hand, refers to the use of living organisms or their products to create new technologies or solve practical problems. It encompasses a wide range of applications, from the production of life-saving drugs using genetically modified bacteria, to the development of biofuels and bioplastics. Biotechnology has revolutionized industry, medicine, and agriculture, offering sustainable solutions to some of our most pressing challenges.

One of the most significant advancements in genetic engineering and biotechnology is in the field of medicine. Scientists are now able to identify and modify specific genes associated with inherited diseases, potentially offering hope for individuals and families affected by genetic disorders. Additionally, biotechnology has led to the development of personalized medicine, where treatments can be tailored to an individual's unique genetic makeup, improving their efficacy and reducing side effects.

In agriculture, genetic engineering has allowed scientists to create crops that are more resistant to pests, diseases, and environmental stressors. This has the potential to increase food production, decrease the use of harmful pesticides, and address issues of global food security. However, it also raises ethical concerns regarding the long-term effects on ecosystems and human health.

While the potential benefits of genetic engineering and biotechnology are vast, it is essential to consider the ethical and social implications they bring. As students interested in genetics, it is crucial to engage in discussions about the responsible and equitable use of these technologies. By critically evaluating the risks and benefits, we can shape the future of genetic engineering and biotechnology in a way that benefits all of humanity.

In conclusion, genetic engineering and biotechnology have the power to transform our world. From medical breakthroughs to sustainable agriculture, these technologies offer immense potential. As students of genetics, it is our responsibility to stay informed, ask questions, and participate in the ethical debates surrounding these advancements. By doing so, we can ensure that genetic engineering and biotechnology contribute to a better, healthier, and more sustainable future.

Ethical Considerations in Genetic Applications

In the field of genetics, advancements in technology have opened up a world of possibilities for understanding and manipulating the human genome. While these developments bring great promise for the future of medicine and human health, they also raise important ethical considerations that must be carefully examined. This subchapter aims to explore the ethical aspects of genetic applications, providing students with a comprehensive understanding of the potential benefits and risks involved.

One of the key ethical issues in genetic applications is privacy and confidentiality. As genetic testing becomes more accessible and affordable, individuals are increasingly faced with decisions about whether to undergo such testing and what information they are willing to share. Students must understand the importance of protecting an individual's genetic information and the potential consequences of its misuse, including discrimination in employment or insurance coverage.

Another ethical concern is the potential for eugenics and designer babies. The ability to select certain genetic traits in embryos raises questions about whether it is morally justifiable to manipulate the genetic makeup of future generations. Students should critically examine the ethical implications of such practices, considering factors such as social inequality and the potential for a loss of genetic diversity.

Furthermore, the use of genetic information in criminal investigations raises ethical dilemmas. While DNA evidence can be a powerful tool in solving crimes and bringing justice, it also raises concerns about privacy and the potential for false accusations. Students should explore the balance between the need for public safety and individual rights, as well as the potential for racial or socio-economic biases in the criminal justice system.

Additionally, the ethical considerations surrounding genetic research involving human subjects must be addressed. Students should understand the importance of informed consent, the potential risks and benefits, and the need for careful oversight and regulation to protect the well-being of research participants.

Lastly, the global implications of genetic applications cannot be ignored. As genetic technologies become more widespread, issues of access and equity arise. Students should consider how these advancements may exacerbate existing social inequalities and the responsibilities of scientists, policymakers, and the global community to ensure that genetic applications are used ethically and for the benefit of all.

In conclusion, the ethical considerations in genetic applications are multifaceted and complex. As students studying genetics, it is crucial to develop a well-rounded understanding of these ethical dilemmas and engage in thoughtful discussions about the implications of genetic advancements. By critically examining these issues, students can contribute to the responsible and ethical use of genetic applications, ensuring that the benefits of genetic research are enjoyed by all while minimizing potential risks and harms.

Chapter 9: Advances in Human Genetics

Genomics and the Human Genome Project

In the rapidly advancing field of genetics, one key area of focus is genomics. Genomics refers to the study of an organism's entire genome, which is the complete set of DNA that makes up an individual. This branch of genetics aims to understand how genes work together to influence the development, functioning, and overall health of an organism.

A groundbreaking initiative that greatly contributed to our understanding of genomics is the Human Genome Project. This international research effort, initiated in 1990, sought to identify and map all the genes in the human genome. The project was a monumental undertaking that involved scientists from around the world collaborating to decipher the building blocks of human life.

The Human Genome Project provided us with a complete sequence of the human genome, which consists of approximately 3 billion base pairs. This sequence serves as a reference for studying genetic variations among individuals, as well as identifying genes associated with specific diseases or traits.

One of the major accomplishments of the Human Genome Project was the development of new technologies that revolutionized the field of genomics. These advancements enabled scientists to sequence DNA faster and more accurately, making genomic research more accessible and affordable. As a result, the field of genomics has expanded rapidly in recent years, with new discoveries and applications being made regularly.

Genomics has numerous applications in various fields, including medicine, agriculture, and forensics. In medicine, genomics plays a crucial role in personalized medicine, where treatments are tailored to an individual's genetic makeup. By analyzing an individual's genome, doctors can predict their susceptibility to certain diseases and develop targeted therapies.

In agriculture, genomics is used to improve crop yield, disease resistance, and nutritional value. By studying the genomes of plants and animals, scientists can identify genes that confer desirable traits and use that knowledge to breed more productive organisms.

In forensics, genomics is utilized to identify individuals based on DNA evidence left at crime scenes. It has revolutionized criminal investigations, leading to more accurate identification and improved justice systems.

As students, understanding genomics and the Human Genome Project is crucial for a career in genetics. The advancements in genomics have opened up new possibilities for research and discovery. Familiarizing yourself with the principles and applications of genomics will provide you with a solid foundation for further studies in this field and prepare you for the exciting future of genetics.

Next-Generation Sequencing

Next-Generation Sequencing: Unveiling the Secrets of the Genetic Code

Welcome to the fascinating world of Next-Generation Sequencing (NGS)! In this subchapter, we will embark on a journey to uncover the mysteries of the genetic code, using cutting-edge technology that has revolutionized the field of genetics. For students passionate about genetics, NGS offers an incredible opportunity to delve into the intricate workings of our DNA and explore the incredible diversity it holds.

NGS is a revolutionary technique that allows scientists to rapidly sequence large amounts of DNA, providing a comprehensive view of an individual's genetic makeup. Unlike its predecessor, Sanger sequencing, NGS enables researchers to decipher the entire human genome in a matter of days, rather than years. This breakthrough has opened up new avenues of research, shedding light on the complex interactions between genes, environment, and disease.

So, how does NGS work? Let's take a closer look. First, DNA samples are extracted from cells and amplified using a process called polymerase chain reaction (PCR). These amplified fragments are then fragmented into smaller pieces and attached to tiny beads, where they undergo the sequencing process. NGS platforms utilize different approaches, such as Illumina sequencing, Ion Torrent sequencing, or Pacific Biosciences sequencing, to determine the order of the nucleotides that make up our DNA.

The immense power of NGS lies in its ability to generate vast amounts of data. These data sets, often referred to as "reads," are then analyzed using sophisticated computational algorithms to assemble the fragmented DNA sequences. By comparing these sequences to a reference genome, scientists can identify genetic variations, such as single-nucleotide polymorphisms (SNPs) or structural variants, which contribute to the diversity of the human population.

NGS has revolutionized various fields of genetics, including medical genetics, evolutionary biology, and personalized medicine. Researchers can now study the genetic basis of diseases, identify potential therapeutic targets, and develop personalized treatments tailored to an individual's unique genetic makeup. Moreover, NGS has provided crucial insights into human evolution, elucidating our shared ancestry and the genetic adaptations that have shaped our species.

In conclusion, Next-Generation Sequencing has ushered in a new era of genetic exploration. For students interested in genetics, NGS offers a unique opportunity to unravel the mysteries of our DNA and understand the world of inherited traits, diseases, and evolution. As we continue to unlock the secrets of the genetic code, NGS will undoubtedly shape the future of genetics and pave the way for groundbreaking discoveries that will benefit humanity for generations to come.

CRISPR-Cas9 and Gene Editing

In recent years, the field of genetics has witnessed a groundbreaking advancement that holds immense potential for the future of humanity – CRISPR-Cas9 gene editing. This revolutionary technology has taken the scientific community by storm and promises to revolutionize the way we understand and manipulate our genetic code.

CRISPR, which stands for Clustered Regularly Interspaced Short Palindromic Repeats, is a naturally occurring system found in bacteria. It acts as a defense mechanism against viral infections by capturing snippets of viral DNA and storing them in the bacterial genome. Cas9, on the other hand, is an enzyme that acts as a molecular scissors, cutting the DNA at specific locations determined by a small RNA molecule. Together, CRISPR-Cas9 allows scientists to precisely edit the genetic material – the DNA – of living organisms.

The implications of CRISPR-Cas9 are far-reaching. For students interested in genetics, this technology opens up a whole new world of possibilities. It allows scientists to study the function of specific genes by selectively disabling or modifying them, providing valuable insights into the genetic basis of various diseases. This understanding could pave the way for the development of new treatments and therapies, potentially leading to the eradication of genetic disorders that have plagued humanity for centuries.

Furthermore, CRISPR-Cas9 holds immense promise in the field of agriculture. Gene editing can be used to enhance crop yields, increase resistance to pests and diseases, and even modify the nutritional content of food. This has the potential to address global food security challenges and ensure a sustainable future for our planet.

However, along with the promises, CRISPR-Cas9 also raises ethical concerns. The ability to manipulate our genetic code brings with it ethical dilemmas regarding the potential misuse or unintended consequences of gene editing. As students studying genetics, it is crucial to be well-informed about the ethical implications and engage in thoughtful discussions surrounding the responsible use of this technology.

In conclusion, CRISPR-Cas9 gene editing represents a monumental leap forward in the field of genetics. As students passionate about genetics, it is essential to stay updated with the latest advancements and understand the potential benefits and ethical considerations associated with this technology. The power to edit and alter our genetic code brings about a responsibility to use this technology ethically and responsibly for the betterment of humanity and our planet.

Precision Medicine

Precision medicine is an emerging field in genetics that holds great promise for the future of healthcare. It is a revolutionary approach that takes into account an individual's unique genetic makeup, environment, and lifestyle to tailor medical treatments specifically to their needs. This subchapter explores the concept of precision medicine and its implications for improving healthcare outcomes.

In the past, medical treatments were often generalized, assuming that what worked for one patient would work for another. However, modern advancements in genetics have revealed that each person's genetic profile is distinct and can significantly impact their response to medications and therapies. Precision medicine seeks to harness this knowledge to provide more effective, personalized treatments for patients.

The foundation of precision medicine lies in understanding the human genome. The human genome consists of all the genetic material present in an individual's cells, which serves as a blueprint for their development, functioning, and overall health. By analyzing an individual's genome, scientists can identify specific genetic variations that may contribute to the development of certain diseases or affect their response to different treatments.

One of the key applications of precision medicine is in cancer treatment. Cancer is a complex disease with various subtypes, each requiring a tailored approach for effective treatment. By analyzing the genetic mutations present in a tumor, oncologists can identify targeted therapies that specifically address the underlying genetic drivers of the cancer. This approach has shown remarkable success in improving survival rates and reducing side effects compared to traditional chemotherapy.

Precision medicine is not limited to cancer treatment; it has the potential to revolutionize many areas of healthcare. For example, by analyzing an individual's genetic predisposition to certain diseases, healthcare providers can implement preventive measures to minimize the risk of disease development. Additionally, precision medicine can help identify patients who are likely to experience adverse reactions to certain medications, enabling doctors to prescribe alternative treatments that are better suited to the individual's genetic makeup.

While precision medicine holds immense promise, it also presents unique challenges. The cost of genetic testing and the interpretation of vast amounts of genetic data are significant hurdles that need to be addressed. Furthermore, ethical considerations surrounding patient privacy and the potential for genetic discrimination must be carefully navigated.

In conclusion, precision medicine is an exciting frontier in genetics that offers the potential to revolutionize healthcare. By considering an individual's unique genetic makeup, environment, and lifestyle, precision medicine aims to provide personalized treatments that are more effective and have fewer side effects. As students in the field of genetics, understanding and exploring the possibilities of precision medicine will be crucial to shaping the future of healthcare.

Future Directions in Human Genetics Research

As students of genetics, it is important to stay updated on the latest advancements and future directions in human genetics research. The field of human genetics is constantly evolving, with new discoveries and technologies revolutionizing our understanding of the genetic basis of human traits and diseases. In this subchapter, we will explore some of the exciting future directions in human genetics research that may shape the field in the coming years.

One promising area of research is the exploration of personalized medicine. With the advancement of technology, it is now possible to sequence an individual's entire genome quickly and at a reasonable cost. This opens up the possibility of tailoring medical treatments to an individual's unique genetic makeup. By identifying genetic variations that may influence an individual's response to certain drugs or susceptibility to diseases, we can optimize treatment plans and improve patient outcomes.

Another future direction in human genetics research is the study of epigenetics. Epigenetics refers to the changes in gene expression that do not involve alterations in the DNA sequence itself. These changes can be influenced by environmental factors, such as diet, stress, and lifestyle choices. Understanding how epigenetic modifications occur and how they influence gene expression can provide valuable insights into the development of complex diseases, such as cancer and cardiovascular diseases.

Advancements in gene editing technologies, such as CRISPR-Cas9, have also opened up new possibilities in human genetics research. This technology enables scientists to precisely edit specific genes in an organism's genome, offering potential treatments for genetic disorders. However, the ethical implications of gene editing are still being debated, and it is essential for students to understand the ethical considerations associated with these technologies.

The field of human genetics is also increasingly focusing on understanding the genetic basis of complex traits and diseases. While single gene disorders are relatively straightforward to study, complex traits and diseases, such as diabetes and autism, involve the interplay of multiple genes and environmental factors. Researchers are using large-scale genetic studies, such as genome-wide association studies (GWAS), to identify the genetic variants associated with these complex traits. This research may help in the development of targeted therapies and preventive measures.

In conclusion, the field of human genetics holds immense potential for future discoveries and advancements. As students of genetics, it is crucial to stay informed about the latest developments and future directions in this field. From personalized medicine to epigenetics and gene editing technologies, the future of human genetics research is filled with exciting possibilities. By keeping abreast of these advancements, students can contribute to the ever-growing body of knowledge in human genetics and help shape the future of healthcare and medicine.

Chapter 10: The Future of Human Genetics

Emerging Technologies in Human Genetics

In recent years, the field of human genetics has witnessed a remarkable transformation due to the advent of emerging technologies. These cutting-edge tools and techniques have revolutionized our understanding of the human genome and hold immense potential for advancements in the field of genetics. This subchapter explores some of the most exciting emerging technologies in human genetics that are shaping the future of genetic research and its applications.

One such technology is CRISPR-Cas9, a powerful gene-editing tool that has garnered significant attention in the scientific community. CRISPR-Cas9 allows scientists to precisely modify specific genes within an organism's DNA, opening up possibilities for curing genetic diseases, creating disease-resistant crops, and even introducing beneficial traits into individuals. Its potential to revolutionize medicine is immense, offering new hope for treating previously incurable genetic disorders.

Next, we delve into the field of personal genomics, which has gained popularity with the advent of direct-to-consumer genetic testing kits. These kits allow individuals to obtain information about their genetic makeup and ancestry with a simple saliva sample. Personal genomics empowers individuals to gain insights into their genetic predisposition to various diseases, enabling them to make informed choices about their health and lifestyle.

Furthermore, the emergence of high-throughput DNA sequencing technologies has greatly accelerated the pace of genetic research. These technologies allow scientists to sequence the entire human genome more rapidly and at a lower cost than ever before. As a result, large-scale genetic studies are now feasible, enabling researchers to identify genetic variations associated with complex diseases and traits.

In addition to these breakthroughs, emerging technologies such as single-cell sequencing and gene expression profiling are unraveling the complexity of human development and disease. Single-cell sequencing provides insights into the diverse cell types within an organism, shedding light on how individual cells contribute to the functioning of tissues and organs. Gene expression profiling allows scientists to measure the activity of thousands of genes simultaneously, aiding in the understanding of gene regulation and the development of targeted therapies.

As students interested in genetics, it is crucial to stay updated with the latest advancements in the field. The technologies discussed in this subchapter represent the forefront of human genetics research and have the potential to transform the way we diagnose, treat, and prevent genetic diseases. By embracing these emerging technologies, we can unlock the secrets of the human genome and pave the way for a healthier future.

Genetic Testing and Privacy

In today's rapidly advancing world of genetics, the field of genetic testing has become increasingly popular and accessible. Genetic testing involves analyzing an individual's DNA to identify potential genetic disorders, predict the risk of developing certain diseases, and even determine ancestry. While the information obtained from genetic testing can be valuable, it raises important concerns about privacy and confidentiality.

Privacy is a fundamental right, and it is crucial to safeguard individuals' genetic information. Genetic testing can reveal sensitive and personal information about an individual's health, including their predisposition to certain diseases. This raises concerns about potential discrimination by insurance companies, employers, and even society at large. For example, an individual with a high risk of developing a certain disease may face difficulties in obtaining insurance coverage or employment opportunities. Therefore, it is necessary to have strict laws and regulations in place to protect individuals from such discrimination.

To address these concerns, various legal frameworks have been established. In the United States, the Genetic Information Nondiscrimination Act (GINA) was passed in 2008 to protect individuals from genetic discrimination. GINA prohibits health insurers and employers from using an individual's genetic information to make decisions about coverage, premiums, or employment. Similarly, the European Union's General Data Protection Regulation (GDPR) provides individuals with control over their personal data, including genetic information, ensuring that it is processed securely and with explicit consent.

Additionally, genetic testing companies also have a responsibility to uphold privacy standards. They must implement robust security measures to protect genetic data from unauthorized access and ensure that individuals have control over how their data is used and shared. It is essential for students to understand the importance of carefully selecting reputable genetic testing companies that prioritize privacy and have transparent policies regarding data usage.

Furthermore, students should be aware of the potential risks associated with sharing genetic information online. Social media platforms and online genealogy databases have become popular tools for individuals to explore their ancestry or connect with distant relatives. However, the sharing of genetic data on these platforms can pose risks to privacy, as data breaches and unauthorized access are always a possibility. Students must be cautious about sharing their genetic information online and should only use secure and trusted platforms.

In conclusion, genetic testing offers valuable insights into our health and ancestry. However, ensuring privacy and protecting genetic information is of utmost importance. Students should be aware of their rights and the legal protections in place to safeguard their genetic data. By making informed choices, selecting reputable testing companies, and being cautious when sharing information online, students can navigate the world of genetic testing while safeguarding their privacy.

Ethical and Social Implications of Genetic Advancements

In recent years, the field of genetics has witnessed tremendous advancements, revolutionizing our understanding of human biology and opening up new possibilities for medical treatments and interventions. However, these advancements also raise important ethical and social considerations that must be carefully examined and addressed. This subchapter explores the ethical and social implications of genetic advancements, aiming to provide students with a comprehensive understanding of the complex issues at play.

One of the most significant ethical concerns revolves around the use of genetic information for discrimination. As our ability to sequence and analyze the human genome improves, the potential for genetic discrimination in areas such as employment, insurance, and education becomes a real possibility. Students need to be aware of the importance of protecting genetic privacy and the potential consequences of genetic discrimination on individuals and society as a whole.

Another ethical dilemma arises from the ability to manipulate and edit genes, known as genetic engineering. While this technology holds promising potential for treating genetic disorders, it also raises ethical questions about the limits of intervention. Students should explore the ethical implications of altering the genetic makeup of future generations and the potential risks associated with genetic engineering.

The social implications of genetic advancements are equally noteworthy. Genetic testing, for example, allows individuals to learn about their likelihood of developing certain diseases. The knowledge gained from such tests can empower individuals to make informed decisions about their health but may also lead to anxiety and psychological distress. It is crucial for students to understand the importance of genetic counseling and support services in helping individuals navigate the emotional and psychological challenges associated with genetic testing.

Furthermore, genetic advancements have the potential to exacerbate existing social inequalities. Access to genetic testing and advanced medical interventions may not be equally distributed, creating disparities in healthcare outcomes. Students should critically examine these social implications and consider ways to address these disparities to ensure that genetic advancements benefit all members of society equitably.

In conclusion, while genetic advancements hold immense potential for improving human health and well-being, they also bring forth a range of ethical and social implications. By exploring these issues, students will gain a deeper understanding of the responsibilities and considerations that come with genetic advancements. It is crucial for students to engage in thoughtful discussions and develop a strong ethical framework to navigate the complexities of genetics in a responsible and socially conscious manner.

Genetics and Personal Identity

Understanding the role of genetics in shaping our personal identity is a fascinating field of study that offers valuable insights into who we are as individuals. In this subchapter, we will explore the intricate relationship between genetics and personal identity, delving into the ways in which our genes impact various aspects of our lives.

Genetics, the study of heredity and the variation of inherited traits, plays a significant role in determining our physical characteristics, such as eye color, hair texture, and height. However, genetics goes beyond mere physical attributes and also influences our susceptibility to certain diseases and our response to medications. By understanding our genetic makeup, we can gain insight into our predispositions and take proactive steps to maintain our health and well-being.

Moreover, genetics plays a crucial role in shaping our personality traits, cognitive abilities, and behavioral tendencies. Researchers have discovered that specific genes influence various aspects of our personality, such as extroversion, impulsivity, and even intelligence. However, it is important to note that genetics is not the sole determinant of our personality; environmental factors and personal experiences also play a significant role.

As students of genetics, it is essential to recognize that while our genes provide a foundation, they do not define us entirely. Our personal identity is a complex interplay between our genetic predispositions and the choices we make throughout our lives. It is the unique combination of our genetic makeup, experiences, cultural influences, and personal beliefs that shape who we are as individuals.

Understanding the impact of genetics on personal identity can also help us navigate ethical and societal challenges. Genetic testing, for example, has become increasingly accessible, raising questions about privacy, discrimination, and the potential misuse of genetic information. By exploring these issues, we can become informed and responsible consumers of genetic technologies.

In conclusion, genetics plays a multifaceted role in shaping our personal identity. It influences our physical attributes, susceptibility to diseases, and even our personality traits. However, it is crucial to recognize that our personal identity is the result of a complex interplay between our genetic makeup and our individual choices and experiences. By understanding the impact of genetics on personal identity, we can gain valuable insights into ourselves and make informed decisions about our health, well-being, and societal participation. As students of genetics, let us embrace the wonders of genetics while remaining mindful of the ethical implications and societal considerations.

Conclusion: The Power and Potential of Human Genetics

As we have explored the fascinating world of human genetics throughout this book, it is clear that the study of genetics holds immense power and potential. From understanding our own unique genetic makeup to unraveling the secrets of inherited diseases, genetics has the ability to transform our lives in incredible ways. In this final chapter, we will summarize the key takeaways and highlight the remarkable opportunities that lie ahead in the field of human genetics.

First and foremost, one of the most important lessons we have learned is that every individual possesses a unique set of genes that contribute to their physical traits, personality, and overall health. Understanding this genetic diversity is crucial in appreciating the rich tapestry of human existence. We are all connected through our shared genetic heritage, yet each of us is a distinct combination of genetic variations.

Furthermore, our exploration of genetic disorders has shed light on the challenges faced by individuals and families affected by these conditions. However, it has also shown the incredible potential for genetic research and therapies to alleviate suffering and improve lives. The advancements in gene therapy and genetic engineering offer hope for a future where many genetic disorders can be effectively treated or even cured.

Moreover, the study of genetics has far-reaching implications beyond individual health. It plays a vital role in fields such as forensics, agriculture, and conservation. By utilizing genetic tools and techniques, scientists can identify criminals, develop disease-resistant crops, and preserve endangered species. These applications demonstrate the positive impact that genetics can have on society as a whole.

As students, you are the future of genetic research and its applications. The knowledge and skills you have gained from this book will equip you to contribute to this exciting field. Whether you choose to become genetic counselors, researchers, or even policy makers, your understanding of human genetics will be invaluable in shaping the future of healthcare and scientific advancements.

In conclusion, the power and potential of human genetics are immense. Studying genetics allows us to comprehend our own uniqueness, address genetic disorders, and make significant strides in various fields. As we embark on this journey, let us remember that with great power comes great responsibility. The ethical considerations surrounding genetic research and applications must always be at the forefront of our minds. By harnessing the power of human genetics responsibly, we can pave the way for a future where genetic diversity is celebrated, genetic disorders are conquered, and the benefits of genetics are enjoyed by all.

Printed in the USA
CPSIA information can be obtained
at www.ICGtesting.com
LVHW010342050624
782319LV00011B/361